YOUR KNOWLEDGE HAS VALUE

- We will publish your bachelor's and master's thesis, essays and papers

- Your own eBook and book - sold worldwide in all relevant shops

- Earn money with each sale

Upload your text at www.GRIN.com and publish for free

Martin Payrhuber

The Changing Meaning of Territorial Borders

A Source of New Kinds of Social and Economic Inequality?

GRIN Verlag

University of Salzburg, Department of Geography and Geology

The Changing Meaning of Territorial Borders

A Source of New Kinds of Social and Economic Inequality?

Martin Payrhuber

Seminar: „Geographies of Inequalities", WS 2013/14

Table of Contents

1 Introduction

Amin (2004: 33) points out that it is odd that the mainstream view of cities and regions is still one of territorial entities, although recent developments have been "transforming cities and regions into sites immersed in global networks of organization and routinely implicated in distant connections and influences" (ibid. Amin). These developments have become known as globalization and were they reason why spatial configurations (e.g. territorial borders) are no longer necessarily territorial or scalar, because "the social, economic, political and cultural inside and outside are constituted through the topologies of actor networks which are becoming increasingly dynamic and varied in spatial constitution" (Amin 2002 cited acc. to ibid. Amin). This paper intends to outline the circumstances and consequences of the development identified by Amin in terms of the creation of new forms of inequality and disparity. The first part deals with definitions in the realm of borders and boundaries, the second part treats the historical, current and future meanings of borders and the third part draws the connection between borders and inequality.

2 Definitions

Before it is possible to deal with the notion of "border", its changing meaning and its implications for inequality, it is necessary to clarify some conceptual details. The English language knows three main terms in this respect, which can sometimes be used interchangeably, but nevertheless have their fine differences when it comes to connotation. These three terms are "border", "boundary", and "frontier". In the following, a list of definitions from dictionaries and scientific literature shall be presented, which is then discussed in more detail:

- **border**
  "a point or limit that indicates where two things become different" (Merriam-Webster Online[1] 2013: s.p.)
- **boundary**
  "a line separating one country or state from another / a boundary between places"
  (Merriam-Webster Online[2] 2013: s.p.)
  "Boundaries, by definition, constitute lines of separation or contact. This may occur in real or virtual space, horizontally between territories, or vertically between groups and/or individuals."
  (Newman and Paasi 1998: 191)

- **frontier**
  "an area in proximity to the border"
  (Newman and Paasi 1998: 189)

Judging from these definitions, it seems that of the three notions, "boundary" is used in the most general way. According to Newman and Paasi (1998: 191), it is not limited to territorial lines of separation and can even refer to virtual space, too, which is probably why this concept is the term used most widely in the literature researched for this thesis. Whenever authors speak of lines of separation between ideas, values, social classes, etc., they tend to use the word "boundary". "Border", on the other hand, virtually always denotes something territorial. The term is mostly used to make it clear that the author speaks of boundaries between places, such as countries, regions, or even supra-state organizations.

While "borders" and "boundaries" basically define the sheer lines of separation, "frontiers" are meant to constitute not only the line itself, but also the land surrounding this line – the so-called "borderland", the area in proximity to the border. The term very often bears a connotation of previously untamed, unsettled, and uncivilized land, probably dating back to the times of the early US settlers moving west in North America. It has since been used most prominently in popular culture in the opening credits to the science fiction series "Star Trek": "Space – the final frontier" are the first words of each episode, used by the authors of the series to illustrate that in the 23rd century, there are no more border(land)s on earth itself, but only in (outer) space.

In the course of this thesis, the term "boundary" shall be used most widely – due to its universality. It can incorporate (territorial) borders as well as frontiers, is not limited to physical constraints and can thus also refer to mental lines of separation. Whenever necessary, however, the more specific terms shall be used to make matters clear and to be as precise as possible.

2.1 Static Boundary Classifications: Antecedence vs. Subsequence

Especially in the larger context of boundaries and inequality, it makes sense to further classify boundaries into antecedent and subsequent ones, as it has an effect on such phenomena as segregation, polarization and exclusion – all of utter importance when it comes to inequality.

Antecedence denotes the creation of a boundary prior to the settling of a region, e.g. the US settlers going west in the 18th and 19th centuries. This means that the boundaries are first (politically) created and only then do people move into the region that is confined by them. Subsequent boundaries, on the other hand, are formed and superimposed upon existing patterns of human settlement, as in borderlines drawn by the prevailing powers after a war (Newman and Paasi 198: 190).

What is important in terms of inequality is that migrants and refugees normally come to antecedent boundaries, i.e. to areas already settled by other people(s), and are usually forced to conform to the pre-existing society, rather than preserving their existing cultural identity. In the case of non-conformance, exclusion is often the result and therefore also inequality in terms of social and economic integration and lastly, also marginalization and poverty (ibid. Newman and Paasi).

2.2 Functional Boundary Classifications: Open vs. Closed

Functional boundary classifications refer to different levels of contact ("openness") or separation ("closure") between peoples and states on both sides of the boundary (ibid. Newman and Paasi). Paasi (2011: 13) sees this functional role of borders as especially crucial for economists who want to analyze their meanings as barriers and thus also the question of how to cross them. The transformation of frontiers from barriers into active spaces has been one of the key issues of European (economic) integration over the past twenty to thirty years and lowering the boundaries between states has been the key motive of European Union spatial politics. A prominent Austrian example is the EU funding of the Burgenland region (now in the phasing-out stage) in order to make this borderland between Austria and Hungary an economically prospering, active space.

3 The Meaning Of Borders

3.1 Artificial and Natural Borders

In order to be able to discuss the change in the meaning of borders that has taken place roughly over the past half-century, one must first have a look at how those boundaries and borders that are changing came into existence in the first place. In a way, the question whether boundaries are natural or artificial (i.e. man-made) can be compared to the chicken and egg dilemma – was the creation of borders a result of the creation of humanity or did borders already exist before mankind? Modern geographers almost completely agree that every border must have been created artificially by mankind, but followers of geographical determinism used to believe in the God-given nature of borderlines until as late as the 19th century. By the beginning of the 20th century, at least a distinction between natural and artificial borders was common, and in the year 1907 a man like Lord Curzon of Kedleston – later to become the British Secretary of State – did not find natural boundaries like mountains, rivers, or forests utterly important any more, but saw them replaced by artificial borders having their "origin in the complex operations of race, language, trade, religion, and war" (Curzon 1907 cited acc. to Redepenning 2005: 169). Curzon was in pursuit of the "good" drawing of borders, by what he meant artificial borders in harmony with natural borders and saw geographical knowledge as most important in this affair. After all, borders cannot be drawn too artificially or else they would lose their authority and legitimacy. Ratzel (1892, cited acc. to Redepenning 2005: 170) supports this theory by stating that the activity of drawing borders is the more successful the more it adheres to natural conditions.

Ever since the times of Ratzel and Curzon, drawing borders and setting up boundaries is therefore seen as an act of humans, which can be done more or less fittingly according to the degree of congruence with natural boundaries, however these might be described, as the human eye sees a high degree of harmony as a result. In this context, Hartshorne (1933, cited acc. to Redepenning 2005: 170) speaks of "boundaries marked in nature".

3.2 The Necessity of Borders

As shall be discussed in chapter 2.4, some scholars argue that there is no need for the existence of borders. The overwhelming majority, however, is of the opinion that a world without borders would be a world without distinctions and without distinctions, the world cannot be visualized. Some scholars argue even that without borders, *nothing at all* can be recognized – only boundaries render human beings capable of handling the world (Luhmann 1994 cited acc. to Redepenning 2005: 175). According to this strain of thought, all the objects and things we use in our daily lives receive their identity only because of a certain distinction we make by drawing some kind of border. Redepenning (2005: 176) agrees with Luhmann and adds that the things we distinguish on a daily basis in the same way become the more powerful the more often we make those distinctions. A spatialization of such distinctions entails an objectification or naturalization thereof, and so gives a special authority to the drawing of borders.

3.3 The Future of Borders: the Changing Meaning

At the end of World War II, the last major "official" change of borderlines of a bigger extent took place. The world was laid out according to the plans of the winning nations and divided into the "East" and the "West" – the Cold War had commenced. Only twelve years later, however, starting with the Treaty of Rome in 1957, which marked the beginning of what we now know as the European Union, a process took off which was to change boundaries again, although in a much slower and subtler way. Borderlines were not simply done away with, but boundaries were altered firstly in the minds of politicians and secondly on a supra-state level.

During the past 15 to 20 years, the major social and political transformations taking place all over the world (the collapse of the rigid Cold War geopolitical divide between West and East) gave the change in the meaning of territorial borders another push. The rapid development of information and communication technologies provided the background for the accelerating globalization – partly generating it and partly illustrating it. The result was the rise of the political and economic importance of regions, as the national borders became less and less important. Nowadays, regions need not necessarily end at a state border, but have become cross-border regions in some cases – a development promoted especially by the European Union in its INTERREG programs. The European integration of many former communist countries in its expansion process starting in 2004 also played an important part in the changing meaning of borders. In general, the functional role of borders and boundaries in Europe as defined in chapter 2.2 shifted from a state of relative closure to a more open one (Paasi 2011: 1ff).

Also the terrorist attack on the World Trade Center in New York City in 2001 can be regarded as change in the meaning of boundaries. The western powers, with the United States of America in the forefront, had to reconsider the meanings of the lines dividing societies, nations, states and even cultural realms. Politicians had to alter their definitions of friend and foe and also the ordinary citizen was forced to incorporate new fears, dividing lines, and insides and outsides into his or her view of the world (Paasi 2011: 2).

3.4 The Complexity of Borders

Because of the substantial changes that have been outlined in the previous chapter, borders and boundaries in general have become quite complex objects of research. Paasi (2011: 5-12) analyzes the background of this complexity and sees the first cause in the perpetually increasing number of state, sub-state and supra-state borders as well as in their changing roles and functions in the globalizing world. More and more cultural groupings are demanding a state of their own, the latest example being the creation of the Republic of South Sudan in the year 2011, and so the current world consists of more than 200 states and more than 300 land borders between them (Paasi 2011: 5). Of these borders, some are relatively closed and strictly guarded (e.g. the borderline between North and South Korea), some are quite open (e.g. the borderlines between the countries belonging to the Schengen zone of the European Union) and others are selectively open, meaning that only one direction can be considered open, while the other direction stays fairly closed, especially for non-citizens of the area within the boundaries (e.g. the external borders of the EU or the US-Mexico border). Adding to this aspect of border complexity is the fact that borders do not stay constant, but on the contrary can be rather volatile, as the case of the EU shows. Political transformations are the reason for some borders becoming softer / more open, and others becoming harder / more closed (Paasi 2011: 5ff).

The second cause for the complexity of borders is identified by Paasi (2011: 8f) in the fact that the meanings attributed to borders are inward-oriented and therefore closely related to the state apparatus, producing such mental constructs as nationalism and territoriality. The latter is an ideological practice that transforms economic success and resources into bounded spaces and creates a "we vs. them" mentality. The outcome is inequality by keeping advantages "inside the borders", while the population outside these borders is not able to benefit. Once again, the European Union is the best example: out of fear for jobs, security and wealth, its citizens might opt for an even harder external borderline of the EU and against a further expansion.

Thirdly, Paasi (2011: 9f) sees the importance of spatial scale as crucial for the complexity of borders. On the one hand, there are hard state borders and on the other relatively soft internal political boundaries. Another dichotomy Paasi identifies is the new regionalism and devolution contrasting with the emergence of ever more supranational borders, the NAFTA (North American Free Trade Agreement) being a rather recent example in the year 1994. Paasi (2011: 10) sees the importance of such unions increasing and believes there is a chance of countering nationalism. The new regionalizations can be regarded as efforts by the nation states to manage and control ethno-nationalist groupings, i.e. to give those groupings a certain autonomy while keeping them from separating completely. For despite the ongoing process of globalization, nation states still seek to keep their relative power – in both directions, towards the "higher", supranational administrative units and the "lower", regional ones.

Lastly, there seems to be another cause for the complexity of borders, according to Paasi (2011: 10ff): several partly competing methodological approaches to border studies. Political economists are first and foremostly interested in the macro-level mobility of capital across borders and the ways of controlling it. Here the focus of research lies upon the changing faces of capitalism, especially the changing global conditions of capital accumulation. Those scholars interested in the power of statecraft will probably be keener on analyzing foreign policy texts and the media in general, as far as the creation of threat and fear supported by borders is concerned. The ethnographic approach again is used by researchers interested in the local narratives that people associate with borders and border crossings in their everyday lives. For them, a key question could be which obstacles people have to face by crossing borders or which advantages people can enjoy by the ever softening borders in some areas.

Due to the matter of borders and boundaries as a research object being such a complex one, the view on their roles can be quite diverse in expert circles, as the following two sub-chapters will demonstrate. Paasi (2011: 17ff) contrasts two opposing perspectives, which he calls "Wishing boundaries away?" and "Borders are everywhere".

3.4.1 Wishing Boundaries Away?

This point of view regards boundaries as "optical illusions" and a hindrance to normative, progressive politics. They are seen as relics of geographical determinism and the cartographic legacy of measuring location on the basis of geographical distance and territorial jurisdiction (Amin 2007 cited acc. to Paasi 2011: 18). These traditions assumed naturalized relations between bounded spaces and certain groups of people, an ontology which is questioned by the representatives of the contemporary "mobility paradigm", who postulate a "complex relationality of places and persons connected through both performances and performativities" (Hannam et al. 2006, cited acc. to Paasi 2011: 19). This theoretical construct in a way also questions the concept of sovereignty, which is sort of shifting from the nation state towards other administrative entities, e.g. supra-national organizations like the EU). Thinking this idea even further, one must also raise the question of the future role of citizenship, which at the moment is still closely connected with borders.

Despite the obviously diminishing relevance of boundaries in the traditional sense, it must be noted that only three percent of the world population live outside their respective states of birth. This small percentage consists mainly of refugees and asylum seekers, who still experience the various faces of exclusion, which seem not to have disappeared just because of a change in the meaning of borders. Rather, a form of "business-class" citizenship for academic guest workers and other high profile foreigners appears to have developed alongside the marginalized parts of the population (Paasi 2011: 21).

3.4.2 Borders Are Everywhere

On the other side of the spectrum are those scholars who suggest that borders have not disappeared, but more or less "become so diffuse that they have transformed whole countries into borderlands" (Balibar 1998 cited acc. to Paasi 2011: 22). This line of thought interprets surveillance technologies like CCTV cameras and increased border control extending beyond the actual borderlines (airports, stations, shopping centers, …) as "a strengthening of border creation in a society" (Rumford 2006, cited acc. to Paasi 2011: 23).

Another strand of the "borders are everywhere" theory sees the creation of emotional borders by nationalism and national identity narratives as crucial. Children learn at school and in their families what the "justified territorial borders" are, as well as what the legitimate and hegemonic national meanings are, which Paasi (2011: 24) calls "emotional landscapes of control".

However, the connotation of borders is by far not homogenous, but context-dependent. A border between one country and another can be quite open (as far as connotation is concerned), whereas the border between the first country and a third can have a very "closed" connotation. Paasi (2011: 24f) mentions the Finnish-Swedish border and the Finnish-Soviet/Russian border as examples and draws the conclusion that there are no automatisms when it comes to the diminishing of borders in the national imagination or socialization. Even though the Cold War has been over for more than two decades now, the Russian-Finnish border is still one that divides and a change in connotation will probably take several generations (Paasi 2011: 25).

4 Borders and Inequality

Living in a world of accelerated globalization and somewhat disappearing (nation state) borders, inequality induced by boundaries whose importance is continually diminishing seems to be sort of a discrepancy at first look. However, as was hinted at in chapter 3.4.2, borders and boundaries may not be vanishing so much as rather shifting towards new areas and transforming themselves into new shapes, as shall be elaborated in the upcoming chapters. Globalization, therefore, does not inevitably signify the "great equalizer", but sometimes even the contrary.

4.1 Inequality Through Migration

The effect caused by borderlines becoming more open and the resulting migration, which did not use to be possible before, can be a rising economic and social inequality in the countries and regions of origin. Even though there were temporary arrangements for migrating workers from Eastern European countries after the expansion of the European Union in the year 2004 until the year 2011, the huge gap between western and eastern European economies led to many citizens of countries such as Poland or Hungary leaving their homes and looking for better paid jobs in the promising west. As a result, large parts of these countries were left without adequate workforce to support the local economy.

An even more drastic example, though somewhat further back in time, is the German reunification. In 1990, the border between the two Germanys was not only done away with symbolically, but in reality, with no temporary arrangements as was the case with the EU expansion in 2004. The weak economy of the former GDR, together with a lack of jobs and widespread poverty according to western standards, caused the emigration of many East Germans to West Germany. A development of intense regional disparity was the result, leaving Germany divided into booming and declining regions.

Figure 1: Migration to and from Eastern Germany

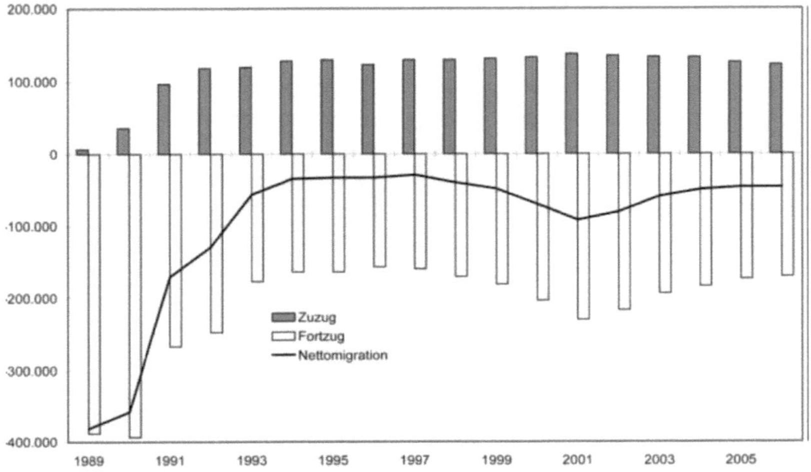

Source: Heydemann 2009: 91

Figure 1 shows the net loss of East German population between the years 1989 and 2005 (the black line), amounting to approximately 2 million people until the present day (Heydemann 2009: 90).

Figure 2: Age-specific migration across German district borders in 1000

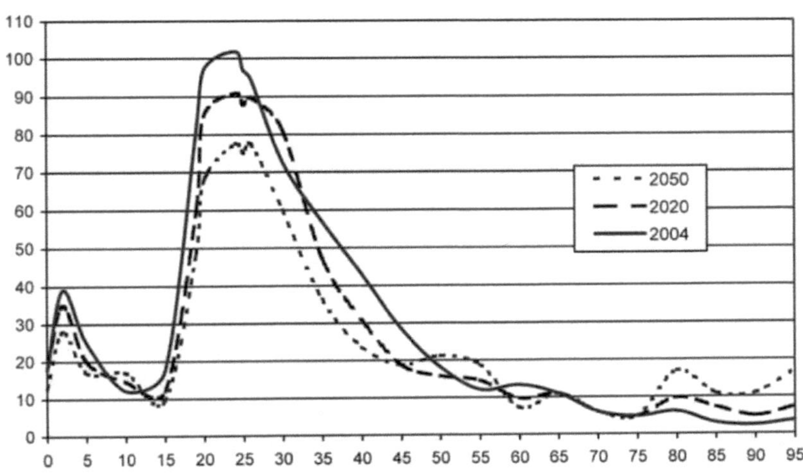

Source: Heydemann 2009: 91

Figure 2 shows another important aspect of German migration: most migrants over district borders (and so assumingly also from Eastern Germany to Western Germany) are between 18 and 30 years of age, i.e. the most fertile population (Heydemann 2009: 92). The result was not only a huge "emptying of East Germany" (Heydemann 2009: 87), but also an increase of the elderly population: from 1991 to 2002, the average population age of the so-called "new Bundesländer" rose from 38.6 to 42.6 – a dramatic development, whose effects can best be explained by Myrdal's "Theory of Cumulative Causation" (Myrdal 1957).

Figure 3: Theory of Cumulative Causation

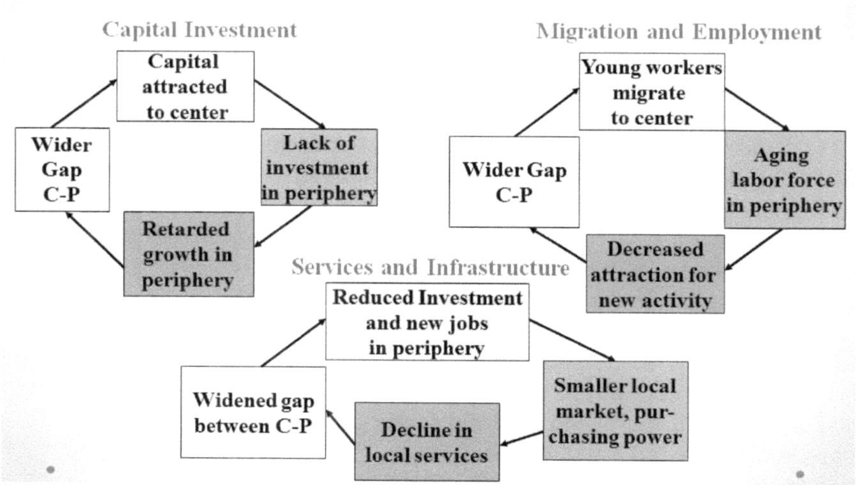

Source: Beyers s.a.: s.p.

Myrdal's model describes how the migration of young workers from the periphery (in the German case East Germany) to the center (i.e. West Germany) leads to an aging labor force in the periphery, decreased attraction for new (economic) activity – as the purchasing power decreases as well – and a gap between the center and the periphery that is continually widening, as the aforementioned factors cause even more people to migrate. The same vicious circle can be observed with services and infrastructure: reduced investment and fewer jobs in the periphery (a result of the decreased attraction from the migration and employment circle) lead to a smaller market and a reduced purchasing power, fewer taxes collected by the state and so a decline in local services. The capital investment circle is similar. In Germany's case, the reunification has thus led to the strongest regional demographic disparities and economic inequalities in Europe during the past 20 years. A possible solution for such a situation is a differentiated promotion and funding of selected regions, for example the so-called "Stadtumbau Ost" (Heydemann 2009: 100). On the hand, it is important not to neglect the regions of the "center" either, as might have been the case with the infrastructure in Western Germany, whose road network seems to be in quite a desolate state compared to the Eastern German network. Without applying the necessary care, the plan can backfire - with center and periphery slowly swapping places.

4.2 Global Cities

While the previous chapter mostly dealt with inequality between different regions, the global city thesis deals (among other aspects) with social polarization and inequality within cities. It was first conceived by Sassen in 1991, who claims that due to globalization the employment structures of cities like New York, London or Tokyo have changed during the second half of the 20th century. This change was characterized by the decline of the manufacturing industry, deindustrialization and the growth of various forms of business and professional services and a particularly strong growth of the FIRE[1] industry (Norgaard 2003: 115). The modifications resulted in a division of the labor market into top earners (managers, CEOs,...) and low earners (employees of service industries like cleaning firms, restaurants, etc.) at the expense of the middle class. These income disparities and the growing informal sector are explicitly mentioned by Sassen (2001, cited ac. to Gerhard 2004: 7) as by-products of the formation of global cities.

Sassen was influenced in the development of her global city thesis by Friedmann, who had already formulated seven "world city hypotheses" in 1986. Although world cities and global cities are not synonymous, they share important details and many world cities are also global cities. Two of Friedmann's hypotheses are essential for the topic of this paper (Friedmann 1986 cited acc. to Gerhard 2004: 5):

- World cities are the points of destination for large numbers of both domestic and/or international migrants.
- World city formation brings into focus the major contradictions of industrial capitalism – among them spatial and class polarization.

It seems that the changing (i.e. decreasing) meaning of territorial borders not only leads to inequality in the regions that are left behind by migrants

---

[1] Finace, Insurance, Real Estate

(as laid out in chapter 4.1), but also in the destinations, which very often are big cities. Arriving workers mostly have only a basic education – if any at all – and are forced to accept the lowest paid jobs, if they even succeed in finding an official workplace - those who fail become part of the informal sector. Apart from this "class polarization", there is also the "spatial polarization", which happens because of various segregation tendencies, both voluntary and involuntary, for example disparities in housing prices or the creation of "ethnic communities".

In this train of thought, Marcuse and van Kempen (2000, cited acc. to Gerhard 2004: 7) speak of a "Quartered City" rather than a "polarized city" (which would imply only two poles), consisting of various enclaves, for example gentrified boroughs, working class neighborhoods or ethnic enclaves. The authors, however, concede that these disparities may have already existed before globalization, but have since been enhanced by the globalization process. The effect of globalization, therefore, really seems to be a paradoxically heterogenizing one, as implied at the beginning of chapter 4.

4.3 Theory of Fragmenting Development

Scholz (2002) picks up the polarization aspect of the global city thesis and speaks of a "fragmenting development" caused by globalization. He blames globalization for not being aimed at consensus, but success, competition and displacement. Thus, its result is a spatial and temporal juxtaposition of integrating and (fragment-like) separating ("fragmenting") processes (Scholz 2002: 7) as shown in figure 4.

Figure 4: Global Fragmenting Development

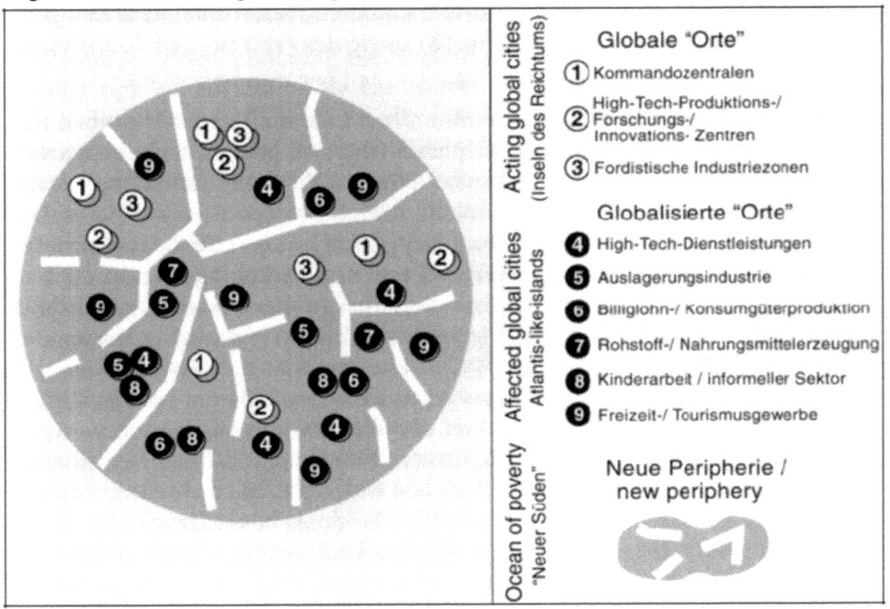

Source: Scholz 2002: 7

Figure 4 illustrates how not countries and states as a whole participate in global competition and its proven benefits, but only certain places (global places / cities and globalized places / cities) and even there, only certain parts of the population – those parts who solely act in their own best interests. The rest of the world is marginalized as the "New Periphery" and serves as a sales market for cheap industrial products, receivers of alms and help in case of catastrophes, deliverers of resources and sources of IT-specialists, professional sportsmen and –women or exotic women. This "New South" is mostly left up to itself as the trifold superfluous "population redundant": Their labor force is not wanted, they are not relevant as consumers and their products are not needed.

Scholz' concept was originally devised to substitute the idea of the Third World countries' development catching up with western standards through the established fact of a "fragmenting development" (Scholz 2002: 11). Part of this fragmenting development is the globally unequal access to the internet, a situation which is expressed in the term "digital divide". According to Scholz (2002: 10), this digital divide is the reality that organizations like the OECD or the World Bank have to face in their hopes of e-commerce representing the way out of poverty for developing countries. There is a chance of using information and communication technologies (ICT) to open up new transnational markets and commercial contacts without having to invest too much into the necessary infrastructure and therefore a chance to increase the economic effectivity. However, a solution must be found to overcome this digital divide, which still separates the globalized world from the "new periphery".

4.4 Gated Communities

A plainly visible consequence of the changed meaning of territorial borders leading to new boundaries are gated communities. The fragmented city landscape also means that different social classes live in close vicinity to each other, which in some societies gave rise to the construction of walls, gates and fences to act as both physical and symbolic markers of exclusion. Pow (2009: 92) has found many documented studies which conclude that private gated communities are most prevalent in places with laissez-faire government regulation and weak state supply of public infrastructures. This seems to explain why gated communities can be found most often in the USA. There, Blakely and Snyder (1999, cited acc. to Pow 2009: 92), have identified three broad categories of gated communities:

- "lifestyle communities", such as retirement villages, golf/leisure community clubs and suburban new towns
- "prestige communities" that symbolize residential distinction as well as to enhance the image and property value in the neighborhood
- "security zone communities" created predominantly out of fear for crime and outsiders

The design principles of gated communities are often antithetical to the basic elements that constitute modern urban life, such as the openness and transparency of urban structures and the free circulation of people and vehicles (Caldeira 2000, cited acc. to Pow 2009: 93). Instead, huge gates and walls often enclose the entire housing estate, leaving the surrounding streets and sidewalks devoid of urban activities and social life (ibid. Pow). On another level, inequality is created through gated communities by the inability of non-gated residents to be free to roam through the neighborhood without being stopped by guards, gates or walls, as well as by increased crime displaced by the presence of a gated community in adjacent neighborhoods (Pow 2009: 94). Therefore, while gated communities may be interpreted as a benevolent landscape, a retreat and even a private haven for middle-class families, they may in fact be considered as landscapes of oppression and segregation for others who were excluded from these privileged spaces (Pow 2009: 96).

4.5 Problems of Inequality and Possible Solutions

So far, this paper has shown how the changing meaning of (territorial) borders can lead to (new kinds of) inequality, implying implicitly that inequality is negative. But what is so bad about inequality? Why should politicians worry about rising disparities in their countries and / or cities? Stegen (2006: 15ff) discusses the consequences of the changes in the labor and housing market, which – according to the global cities thesis (see chapter 4.2) – have been caused by globalization. He argues that while the socially, culturally, politically, and economically more active and wealthier parts of the city populations have been moving from the centers to the suburbs for over a quarter of a century now, the marginalized native citizens, who are disconnected from the labor market, as well the discriminated migrants concentrate more and more in the boroughs of the least quality. Inhabitants of theses boroughs able to afford it also move to better boroughs. This selective mobility leads to a further isolation of the residents of socially segregated boroughs and enhances the downward trend of such areas. The rising number of people moving to these quarters intensifies the competition for jobs and housing space, as a rising number of people looking for jobs and apartments are in opposition to a decreasing number of jobs and affordable housing space (Häußermann 2000 cited acc. to Stegen 2006: 16). Thus, there are parallel spatial and social polarization processes, which finally lead to the creation of so-called "islands of poverty".

The missing access to the education and labor market as well as the spatial segregation from the other urban boroughs is the reason for the increasing loss of connection to the society. Unemployment money is the traditional way of dealing with this situation on the politicians' behalf. This system is supposed to allow every person to lead a humane life until being able to improve the situation of his or her own accord as soon as possible. To this end, however, it is of essential importance that especially socially vulnerable groups possess local contacts and small-scale social networks, because these represent informal potentials that - while not directly leading to new jobs - are the basis for the development of self-supporting structures with a local economy of its own. It is unfortunate that exactly these "forms of capital", which are much easier to generate for the poorer parts of the population, are destroyed in the course of political urban modernization processes (refurbishment, vacation, relocation, demolition) (Stegen 2006: 16f).

According to Stegen (2006: 17f), the solution to the problem of marginalization and socio-spatial polarization lies in concepts of social balance, which support, promote, and cross-link the aforementioned "soft" forms of capital. The author gets more specific in stating that it is important for city politics to re-link the labor and housing markets, as both are closely connected with social issues and adds that the interweaving of economic, social, cultural, and city politics is crucial. The focus, however, should not be the individual-related welfare case, but the area-related social space. What seems to be put down on paper so easily is unfortunately not implemented that easily, as also Stegen (2006: 20f) concedes. The problematic implementation is discussed by the author at length (cf. Stegen 2006: 21ff), but would lead too far in the course of this paper.

5 Conclusion

This thesis has set out to explore how a change in the meaning of territorial borders can lead to new kinds of social and economic inequality. While the concept of *change* has been dealt with extensively (esp. in chapter 3), as well as the way inequality is created thereby (esp. in chapter 4), what is left is to answer is the question what is new about these kinds of social and economic inequalities.

In a way, the inequalities of the pre-Cold-War era were less visible. The borders were synonymous with barriers and more or less closed, at least concerning the former communist-ruled countries. Information and communication technologies were still analogue (and therefore slow), one had to rely on printed newspapers to inform oneself about the activities abroad. The inequality was between entire countries or even political systems, but within these countries and systems, the population was more or less socially and economically homogenous (apart from political leaders). Because there was little contact between the systems, the inequality was not an issue. With the advent of globalization and all of its byproducts, however, a new world order was inevitable and with it came the already discussed change in the meaning of borders, leading to inequalities being first of all much more visible and secondly existing within a much smaller area (especially in cities, as discussed in chapters 4.2, 4.3, and 4.4).

Today, the wealthy and the poor may live next to each other – some societies see this as a problem and build walls and fences between them. On the other hand, it can also be seen as a challenge to tear down even invisible boundaries between different parts of the population, instead of erecting physical new ones. Germany is on a good way with projects like "Die Soziale Stadt" (cf. Stegen 2006), even though there might still be flaws, as Stegen (ibid.) has pointed out in his analysis, and it can (and should) serve as an example for societies like the USA, which are rather dominated by the (voluntary) segregation concept.

6 Literature

Amin, A. (2004): Regions unbound: Towards a new politics of place. In: Geografiska Annaler 86 B (1), 33 – 44.

Beyers, W. (s.a.): Backwash circuits. <http://faculty.washington.edu/beyers/Discussion%20Week%203.ppt> (access: 12/28/13)

Gerhard, U. (2004): Global Cities – Anmerkungen zu einem aktuellen Forschungsfeld. In: Geographische Rundschau 56 (4), 4 – 10.

Heydemann, G. (2009): „Blühende Landschaften" oder entvölkerte Landkreise? Die neuen Bundesländer zwischen Wachstums – und Schrumpfungsprozessen. In: Totalitarismus und Demokratie 6, 87–100.

Merriam-Webster Online[1] (2013): Border. <http://www.merriam-webster.com/dictionary/border> (access: 12/28/13)

Merriam-Webster Online[2] (2013): Boundary. <http://www.merriam-webster.com/dictionary/boundary> (access: 12/28/13)

Myrdal, G. (1957). Economic theory and underdeveloped regions. London: Methuen.

Newman, D. and Anssi Paasi (1998): Fences and neighbours in the postmodern world: boundary narratives in political geography. In: Progress in Human Geography 22,2, 186 – 207.

Norgaard, H. (2003): The global city thesis – social polarization and changes in the distribution of wages. Geografiska Annaler 85 B (2), 103–119.

Paasi, A. (2011): A Border Theory: An unattainable dream or a realistic aim for border scholars? <https://www.academia.edu/attachments/31022427/download_file> (access: 12/28/13)

Pow, C-P. (2009): 'Good' and 'real' places: a geographical-moral critique of territorial place-making. In: Geografiska Annaler 91 B (2), 91 – 105.

Redepenning, M. (2005): Über die Unvermeidlichkeit von Grenzziehungen. In: Berichte zur deutschen Landeskunde 79 (2/3), 167-177.

Scholz, F. (2002): Die Theorie der „fragmentierenden Entwicklung". In: Geographische Rundschau 54 (10), 6 – 11.

Stegen, R. (2006): Die soziale Stadt. Quartiersentwicklung zwischen Städtebauförderung, integrierter Stadtpolitik und Bewohnerinteressen. Berlin: Lit Verlag.